applied surface engineering

February 24

2014

Surface engineering is the branch of science that specifically deals with the numerous methodologies, used in obtaining the desired surface requirements of technological components. In other words, it is a sub-discipline of technology that studies the surface of solid matter. The application of surface engineering has lead to the production of better technological products

By : RAFIS ISMAIL

Contents

1.0 INTRODUCTION 3
2.0 APPLICATIONS 7
   2.1. Application of Surface Engineering in Manufacturing 7
   2.2. Application of Surface Engineering in Transportation Business 8
   2.3. Application of Surface Engineering in Civil designs 8
   2.4. Application of Surface Engineering in Aerospace Industry 8
   2.5. Application of Surface Engineering in Sport 9
3.0. PROCESSES 10
   3.1. Thin film coatings 10
   3.2. Sputtering Deposition Process 11
   3.3. Thermal spraying 12
   3.4. Surface Modification 14
4.0 CHARACTERIZATION 15
   CONCLUSION 16
Bibliography 18

1.0 INTRODUCTION

According to Halling [1], surface engineering is the branch of science that specifically deals with the numerous methodologies, used in obtaining the desired surface requirements of technological components. In other words, it is a sub-discipline of technology that studies the surface of solid matter. The application of surface engineering has lead to the production of better technological products. Figure 1 explains how this is normally accomplished

Figure 1: Role of surface engineering in the process of manufacture of a material product

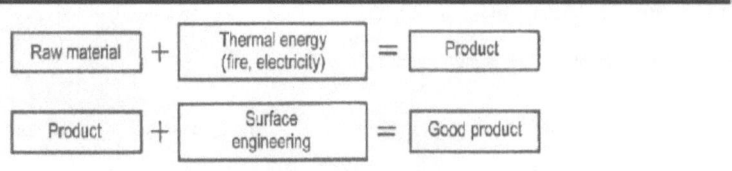

Source: Frainger and Blunt [2]

Basically, the origin of surface and material technology is almost as old as mankind. However, the first known technology that belongs to surface engineering is the heat treatment [3] [4]. This includes both the diffusion (carburizing, nitriding) and non-diffusion (hardening) heat process. These technologies were widely used by the ancient Egyptians, Indian and Chinese, in manufacturing the crude implements that were then in use. For instance, as far back as 2000-1500 BC, the ancient Egyptians and Indian were known to have tried to enhance their iron implements by certain elements into its surface, through the carburizing methods. The heat treatment of steel, which was widely believed to have originated in Damascus, was perfected by the Arabs in 1400 AD and introduced into Europe [4]. But before then, crude steel implements were widely in used in ancient Japan and Roman Empire.

A major breakthrough in material technology was recorded in 1789, with the discovery of the so-called "animal electricity" by L. Galvani (5). This discovery has generally been accepted by many scholars, as the beginning of the development of technologies within surface engineering, which make use of the flow of current, through an electrolyte. Today, this process which is commonly known as electroplating (galvanostegy) is widely employed in depositing valuable elements on the surface on metals and even non-metals. The scientists that actually laid the foundation for electroplating, as it is known today were an Italian scientist named A. Volta (1801) and a Russian Scientist known as B.S. Jacobi. Today, electroplating (galvanostegy) is one of the foremost technologies, being used in Surface Engineering.

The adoption of heat energy in softening or re-melting, consequently lead to the development of welding technology and thermal spraying. Thermal spraying is a more convenient method for depositing coatings on metal and non-metal surfaces. Basically, it involves the use of a heat source to soften or re-melt the material to be sprayed on the surfaces. The ordinary fame, was the first natural heat source to be used in welding, while the forged flame is the oldest artificially heat source. Coincidentally, forge flame served as the basic heat source, for heat several years.

In 1905, the evolution of thermal spraying assumed a very interesting dimension, following the development of the hydrogen-oxygen welding torch. Without wasting much time, the electric arc became the main heat source, for all heat treatment. It was later overtaken by the plasmaburner [6].

The spray gun was later developed by A. Scoop in Switzerland [7]. The great scientist accomplished the feat by atomizing molten metal with the help of a high velocity of stream of gas and placing metal sheet in the way of the stream [7]. The spray gun developed by A. Scoop and his team, consist of pot, flame, wire, powder and arc.

During the second half of the 20th century, the development of the thermal spaying was intensified. This was prompted by the practical utilization of the plasma controlled atmosphere, vacuum and super-sonic spraying [8]. A major breakthrough was recorded in 1955, when R.M. Poorman achieved the first ever utilization of the energy of detonation of explosive material. In his own case, the energy was utilized and used in the deposition of coating on a metallic substrate [8].

During the first half of the 20th century, several researches aimed at understanding the interaction of electron beam with materials are also conducted. Some techniques were consequently developed. First was the conceptualization of the fundamental of shot peening. This was followed by the practical implementation of glow discharge in gases at partial pressure. The researches also led to the conceptualization and subsequent development of ion implantation in metals and metalloid from the gas phase.

The used of forced emission in the amplification of microwaves was developed during the 1950s and 1960s. During that same period, the laser technology was developed and successfully implemented. The practical utilization of ion implantation was accomplished, along with methods of and chemical vapor deposition from the gas phase. . Finally, besides plasma, detonation gun spraying came to be used.

It's very important to note that rapid development and implementation of technical methods, techniques and technologies, that are useable in surface engineering were recorded in the 1960s and afterwards. These latter technologies make use of concentrated or directed beam of coherent photon beam, high power density, solar energy, infrared radiation, plasma and ion beam.

Today, the newest methods that are being used in surface engineering are based on the most recent discoveries in science and technology. These include the beams of electrons, ions and photon. Even though these methods are highly efficient and specialized, the accompanying high cost is one major issue of great concern.

In this report, we are going to analyze the different real surface engineering applications, the major processes involved and the necessity for characterization and performance evaluation in surface engineering.

2.0 APPLICATIONS

In today's modern world, surface engineering is actively being applied in various aspects of human technology. These include electronics technology, sports technology, food industries, mining industries, chemical industries, petroleum industries and transport/aeronautic industries. Basically, there are two main categories of surface engineering methods namely: surface coatings and surface modification.

Surface coating is currently being applied in specialized areas such as: thermal sprayed coatings in biomedical/orthopaedics (e.g. hydroxylapatite), sports (swimming, horses hooves, golfing, sport wear), dentistry, bronze applications, cancer therapy and art industry (e.g. glass colouring and enameling). The typical coating materials that are used in this case are: alloys, metals, thermal barrier coatings, nitrides, diamond like carbon, decorative coatings, etc.

2.1. Application of Surface Engineering in Manufacturing

Surfacing coating is also being applied in the manufacturing of tools and implements. For instance, cutting appliances are being currently being coated with thin wear-resistant materials (2). This helps to boost the efficiency and reliability of such specialized appliances. Some common examples of materials that are being used as coated materials include: Titanium, AlSi alloy etc. Normally, the thickness of the coatings can be of several microns. These materials are also resistant to tear, wear, diffusion and friction.

2.2. Application of Surface Engineering in Transportation Business

As already pointed out earlier, surface engineering is being applied in the transportation industry, where it's being applied in the power units, vehicle components and fixed permanent structures. In each of these categories, surface coating is actively being employed to get rid of tear, wear and friction. In the automobile industries, such vehicular components like, suspension and brakes, are being coated with thermally sprayed coatings to improve wear resistance. Similarly, epoxy-based polymer coatings are applied to exposed areas such as wheel arches and bumpers. The abrasion and corrosion resistance of motor body parts can also be boosted by surface coating.

2.3. Application of Surface Engineering in Civil designs

In civil engineering, surface coating is being applied to obtain desirable qualities. The corrosion rate of the metallic materials used in the construction of bridges and oil rigs can be greatly minimized through surface coating. The same technology can also be used to combat sand abrasion problems.

2.4. Application of Surface Engineering in Aerospace Industry

Surface engineering is also being applied in the aerospace industry. For instance, the temperature strength, corrosion-resisting ability and bearing properties of surface coated gas turbine engine can be improved with surface coating. The parts of the aircraft that are susceptible to atmospheric corrosion can be protected by thermally sprayed polymer coating. Surface modification can also be used to improve the performance of certain components of an aircraft. For instance, in light combat aircraft, the slat track (a component of the landing gear) is formerly designed out of maraging steel. However, there was the need to improve the wear resistance of this component. This was accomplished through surface modification. Using the conventional hard chromium plating may give rise to the problem of machining. Thus, plasma nitriding surface modification was developed and implemented.

2.5. Application of Surface Engineering in Sport

Surface engineering has also been applied in the field of sport. To start with, the engine parts of racing cars have been greatly optimized through surface engineering of titanium alloys, which are highly bio-compatible, resistant to corrosion and has high strength/weight ratio [9]. Another common application of surface engineering is in the golf clubs, where the technology is being used to control friction through surface modification. In this case, the main coating materials that are being used include: TiN, Carbide coatings shot peening and surface shape design. Surface coating and modification are also applied in snow ski designs, curling, cycling, soccer boots and protective clothing. Reducing or increasing friction in the competitive sports can be the difference between winning and losing.

3.0. PROCESSES

There are lots of techniques and processes that generally applied in surface coating. These processes can be placed into the following major groups namely: Thin Film Coating, Sputtering Deposition Process, Thermal spraying and surface modification.

3.1. Thin film coatings

This process is currently being applied in the development of high quality sports equipment, optics, micro-mechanics, bio-medical, magnetic etc. Several deposition methods are currently used in thin film coating. The two commonest methods are Physical Vapor Deposition (PVD) and Chemical Vapor Deposition (CVD). Another advanced method is the Plasma enhanced CVD (PECVD).

As already stated in the last section, the abrasion and corrosion-resisting ability of certain technological components are greatly enhanced by surface coating. These thin film coatings are usually deposited by using PVD, CVD and the advanced Plasma enhanced CVD (PECVD).

Basically, the process of PVD involves the generation of coating vapors. These coating vapors can either be generated by ejection of atoms from a solid source (with the solid source being bombarded by an ionized gas) or through evaporation from a molten source. The coating vapors that are generated can either be ionized or partially ionized and then released. Alternatively, the coating vapor may be allowed to leave as a stream of neutral atoms in a vacuum. Whichever is the case, the coating vapors will eventually be deposited on substrate meant for it. Generally, the thinness of the films that are usually achieved with PVD techniques is within the range of 0.1 μm to 0.1 mm [2].

The CVD coating technique has a broader application. The coating materials that are normally used here include TiN and TiC. The final products that are obtained are capable of withstanding abrasion and corrosion to a greater extent. The most important step in the CVD process is the chemical reaction that takes place between the source gases in a chamber. This reaction produces a solid phase material that is eventually condensed on the substrate surfaces.

In any CVD process, the chemical reaction taking place in the chamber involves three significant steps. These are:

Step 1: Production of the volatile carrier. This can be a chemical compound. A very good example is nickel carbonyl;

Step 2: Transportation of the gas to the decomposition gas. This gas must be transportation to the decomposition site, without undergoing any decomposition.

Step 3: The occurrence of the chemical that will produce the coating material to be used on the substrate.

Unlike the PVD, the CVD can be used for thin films that are in range of 0.1 μm to 0.1 mm as well as those above 1 mm [2].

3.2. Sputtering Deposition Process

In this process, the coating material is subjected to high energy particles, with a resultant dislodgement and ejection of particles of the coating materials. The high energy particles can either be energetic neutrals, positive ions or even a species of the coating material. The dislodged and ejected coating materials, normally known as the target is always in atomic form.

In the sputtering deposition process, the fundamental processes are glow discharge and ion beam. In the first case, the target is placed in a vacuum chamber that has been evacuated to 10^{-7} to 10^{-5} torr. Once the target is in there, the vacuum chamber is then backfilled with a working gas (e.g inert gases, like Ar). The backfilling of the vacuum chamber increased the pressure to the range of 5×10^{-3} to 10^{-1} torr, which is enough to sustain a plasma discharge. The next step is the application of a negative bias (0.5 to 5 kV for d.c. device) to the target. The essence is to make sure, that the target is bombarded by positive ions. The ions generated by this bombardment bring about, the ejection of the target material (sputtering).

The inability to independently control the current density and voltage of the target, limit the reliability of the glow-discharge sputtering technology. The two aforementioned quantities can only be controlled by varying the working gas pressure.

The defects of the glow-discharge sputtering technology are rectified by the magnetically assisted glow-discharge sputtering process, otherwise known as the magnetron sputtering. This enhanced method provides: higher deposition rates, lager deposition areas and lower substrate heating. This has expanded the practical applications of the sputtering deposition process as wider range of metal carbides, nitrides and oxides are currently being used. Some good examples include: TiN, CrN, TiC, DLC, TiAlN, MoS2.

3.3. Thermal spraying

Thermal spraying is another coating process that has been widely used in the surface coating procedures that were discussed in the last section. It's actually one of the most versatile techniques that is available for applying protective coatings. The widely used coating materials are metals, polymers and ceramics. The three basic steps of the thermal spraying coating deposition process are preheating, depositing and fusing. The simplest thermal spraying method is the flame spraying, which exist in two common forms, namely: wire and powder.

Figure 2: An example of a typical powder flame spray process

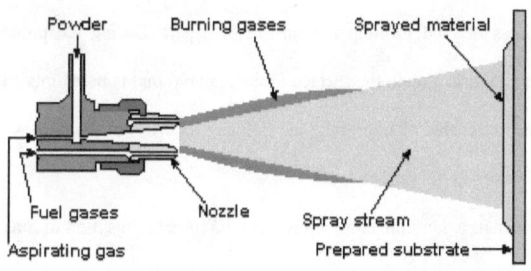

Source: Kennedy et al (10)

One of the commercially available powder flame spray equipment is the Eutectic Castolin's Superjet Eutalloy torch [10]. In this equipment, the oxygen and acetylene are directed separately by different needle valves. The essence is to obtain a precise flame adjustment. Whereas the oxygen passes through an injector that pulls in powder into the system, the acetylene gas channeled separately to a mixer assembly. The two gases are mixed up in the mixer assembly, together with the powder. From here, the resultant mixture will then be channeled through the system, to the spray tip and into the flame. Today, the Eutectic Castolin's Superjet Eutalloy torch is being used in joining and dressing up of metallic components, surfacing or overlaying operations, mould reclamation or finish machining errors of tool and die work.

3.4. Surface Modification

The two main surface modification techniques that are currently being used in Surface engineering are: shot peening and ion implantation. Kennedy et al [10] defined shot peening as a modification process of cold forming the surface of a part. During the process, a stream of hardened steel shot is forced upon the surface of the material that is being modified. The process is nonabrasive and is capable of improving the fatigue properties of the part by the introduction of compression stresses in the surface layer.

Just like the name, ion implantation involves the forceful injection of ions into the surface region of the substrate. This method has been discovered to be very effective in improving the hardness, wear and corrosion resisting ability as well as the fatigue properties of the materials.

4.0 CHARACTERIZATION

Materials Characterization is a multi-disciplinary process that is basically used to determine those features of structure and composition of materials that are relevant to specific preparation, use, study or properties and meet the requirements for the production of a particular material [11]. Generally, the main objectives of material characterization are to identify and quantify the constituents that made up material and consequently determine its reliability, efficiency and applicability.

In surface engineering, several methods are being employed in the characterization process. These methods include: Scanning Electron Microscopy (SEM), Optics Characterization, Atomic Force Microscopy (AFM), Transmission Electron Microscopy (TEM) etc. Each of these methods is capable of analyzing the microstructure of materials, thereby determining its reliability, efficiency and applicability in the field under study.

Basically, the Scanning Electron Microscopy imaged the microstructure of the material under study, with the help of high-resolution field emission scanning electron microscopy (FE-SEM). The exact areas of the materials that are imaged include the surfaces and cross sections area, while the images obtained normally has a resolution in the micro- and nanometer range and a high depth of focus. Again, the three-dimensional effect of the microscopy makes it possible to obtain images with high information content. An extensive characterization of the materials can be accomplished by using energy-dispersive X-ray spectroscopy [12].

Another great surface characterization technique is the Atomic force microscopy (AFM). In this case, the interaction forces that exist between sample and the tip are used to map surface topography on the nanometer scale. The use of the appropriate devices can also enable analysts to use AFM to determine the right material properties, through that same interaction force that exist between the tip and sample. A very good example is the WITec's Mercury 100 AFM, which was fortified with the Digital Pulsed Force Mode (DPFM). The Digital Pulsed Force Mode specifically helped to facilitate the imaging of all those surface properties that can be obtained from force distance and pulsed force curves [13].

CONCLUSION

The practical application of technologies relating to surface engineering, has led to the production of high quality engineering products. This is not surprisingly as the serviceability and life cycles of engineering products are greatly affected by the surface characteristics of the materials used in the design. The natural severity of engineering environments, has make it very necessary to fully employ surface coating process, in the production of high quality components that have improved service life. This means that Surface Engineering technology is the right source of solutions for extreme applications. Indeed, the technology supply real added value and thus profits, to the relevant technologies.

Currently, there are numerous opportunities for researchers that are specialists in this field of discipline. The future is very bright for the technology as there is the tendency to generate huge profits by developing newer and more reliable coating materials. In the future, surface engineering is expected to generate several products that should be used in the sports industry. The reduction in equipment costs coupled with increased technical skills have make future development to become more economically feasible.

Surface engineering is one area of technology that has attracted scientific expertise from different discipline, such as physics, biology, engineering, human medicine etc. This integrated knowledge and expertise is exactly what is required in the development and consequent implementation of dated technology. These advances, supported by environmental, health and safety, nanotechnology and designs will surely lead to greater applications in the coming years.

Bibliography

1. **Halling, J.** *Introduction: Recent Development in Surface Coating and Modification Processes.* London : MEP, 1985.

2. **Frainger, S. and Blunt, J.** *Engineering Coatings – Design and Application.* Abington : Abington, Publishing, 1998. 2nd Edition.

3. **Burakowski, T.** *Heat treatment in China.* 72, s.l. : Metaloznawstwo i Obróbka Cieplna (Metallurgy and Heat Treatment), 1984, , Vol. 71. 3-9.

4. **Burakowski, T. and Marczak, R.** *The service-generated surface layer and its investigation(in Polish).* 103, s.l. : Zagadnienia Eksploatacji Maszyn (Problems of Utilization of Machines), Polish Acad-emy of Sciences - Committeee for Machine Building, 1995, Vol. 30. 327-337.

5. **Bell, T.** *Surface engineering, past, present and future.* 1, s.l. : Surface Engineering, 1990, Vol. 6. 31-40.

6. **Linnik, V.A., and Peskev, P.Yu.** *Contemporary technology of thermal gas deposition of coatings.* Moscow : Publ. Masinostroyenye, 1985.

7. **Kirner, K.** *Geschicht des Termischen Spritzens Entwicklung zu den verschiedenen High-Tech-Verfahren.* Tagungs Unterlagen : 3 Kolloquium on Hochgeschwindigkeits-Flammspritzen, Ingolstadt. , 1994.

8. **Pawlowski, L.** *The science and engineering of thermal spray coatings.* Chichester : John Wiley & Sons, 1995.

9. **Bell, T., and Dong, H.** *Surface engineered titanium: material of the 21st century.* s.l. : Foresight in surface Engineering, Surface Engineering Committee of the Institute of Materials, 2000.

10. **Kennedy D., Xue Y., Mihaylova E** *Current and Future Applications of Surface Engineering..* s.l. : The Engineers Journal (Technical), 2005, Vol. 59. 287-292.

11. **Pizzo.** *Materials Characterization.* s.l. : http://www.engr.sjsu.edu/WofMatE/Mat%27sChar.htm, 2012.

12. **Schulz, U.** *Analysis of coatings and surfaces using scanning electron microscopy (SEM).* s.l. : http://www.iof.fraunhofer.de/en/competences/messverfahren-und-

charakterisierung/oberflaechen-schichtcharakterisierung/analysis_of_coatingsandsurfacesusingscanningelectronmicroscopyse.html, 2013.

13. **AZoNano.** *Surface Characterization Using Atomic Force Microscopy (AFM): New Solutions for Optical and Scanning Probe Microscopy.* s.l. : http://www.azonano.com/ads/abmc.aspx?b=7027 , 2012.

Note:

Note:

Note:

Note:

Note:

www.ingramcontent.com/pod-product-compliance
Lightning Source LLC
Chambersburg PA
CBHW021001180526
45163CB00006B/2454